Contents

Introduction

To assist U.S. manufacturers, exporters and individuals living or traveling abroad, this publication lists the characteristics of electric current available and the type of attachment plugs used in most countries. It is an update of a similar handbook published in 1991. The tables indicate the type of current (alternating or direct current), number of phases, frequency (hertz), and voltage, as well as the stability of the frequency and the number of wires to a commercial or residential installation. This information pertains to domestic and commercial service only. It does not apply to special commercial installations involving relatively high voltage requirements or to industrial installations.

For most countries listed here, two nominal voltages are given. The lower voltages are used primarily for lighting and smaller appliances, while the higher voltages are used primarily for air conditioners, heating, and other large appliances. Travelers planning to use or ship appliances abroad should acquaint themselves with the characteristics of the electric supply available in the area in which the appliance is to be used. In some cases, a transformer may be used to correct the voltage. However, if the appliance requires exact timing or speed and if the frequency of the foreign electricity supply differs from the one the appliance was designed for, it is advisable to use an appliance designed for the foreign frequency since auxiliary equipment to change frequency is bulky and expensive. Some foreign hotels have circuits providing approximately 120 volts which allow guests to use electric shavers and other low-wattage U.S. appliances.

The information presented here was compiled over a period of months from a large number of sources. Consequently, there is some possibility of errors or omissions for which the Department of Commerce cannot assume responsibility. In addition, this information should not be taken as final in the case of industrial or highly specialized commercial installations. It would be impossible for the Department to maintain complete data on every foreign industrial installation. For special equipment for commercial use or heavy equipment for industrial use, the current characteristics for the area of installation should be obtained from the end user.

The 1998 edition was prepared by the Trade Development unit in the International Trade Administration, U.S. Department of Commerce. The information was compiled by John J. Bodson, industry specialist in the Office of Energy, Infrastructure, and Machinery. Editing, desktop publishing, and production were done by Rebecca Krafft, of the Trade Information Division of the Office of Trade and Economic Analysis.

The cooperation of various government and private agencies in providing data is gratefully acknowledged. Special thanks go to the U.S. Foreign and Commercial Service of the U.S. Department of Commerce and the Foreign Service of the U.S. Department of State.

Questions about the content of this publication should be directed to the Energy Division in the Office of Basic Industries, (202) 482-4931.

Key to Terms

Type of current—a.c. indicates alternating current; d.c. indicates direct current.

Frequency—Shown in number of hertz (cycles per second). Note that even if voltages are similar, a 60-hertz U.S. clock or tape recorder will not function properly on 50 hertz current.

Number of phases—1 and 3 are the conventional phases that may be available.

Nominal voltage—The term nominal voltage is used to denote the reported voltage in use in the majority of residential and commercial establishments in the country or city. Direct current nominal voltages are 110/220 and 120/240. The lower voltage is always 1/2 of the higher voltage. On a direct current installation, the lower voltage requires two wires while the higher voltage requires three wires.

Alternating current is normally distributed either through 3 phase wye ("star") or delta ("triangle"), 4-wire secondary distribution systems. In the wye or star distribution system the nominal voltage examples are 120/208, 127/220, 220/380, and 230/400. The higher voltage is 1.732 (the square root of 3) times the lower voltage. In a delta or triangle system, 110/220 and 230/460 are examples of nominal voltages. The higher voltage is always double the lower voltage. The higher voltage is obtained by using 2 or 3 phase wires and the neutral wire while the lower voltage is the voltage between the neutral wire and one phase wire. The higher voltage may be single or 3 phase while the lower voltage is always single phase and used primarily for lighting and for small appliances.

Type of attachment plug in use—Attachment plugs used throughout the world come in various forms, dimensions and configurations too numerous to describe in this report. This report does, however, attempt to point out the basic and most commonly used types of plugs by country. Adapters may be purchased to change from the American plug type to other types.

Number of wires to the consumer—The number of wires which may be used by the consumer is shown. Normally, a single phase, 220/380 volt system or 127/220 system will have two wires if only the lower voltage is available (one phase wire and the neutral). It will have three wires if both the higher and lower voltages are available (two phase wires and the neutral) and where three phase motors will be used, four wires will be available for the higher voltage (the three phase wires and the neutral wire).

Frequency stability—"Yes" indicates that the frequency is stable and that service interruptions are rare.

Type of Plug by Country

Country	Plug Type	Country	Plug Type
Afghanistan	D	Canary Islands	C, E
Albania	C	Cape Verde, Rep. of	C, F
Algeria	C, F	Cayman Islands	A, B
Angola	C	Central African Republic.	C, E
Argentina	C, I	Chad	E
Australia	I	Chile	C, F, L
Austria	C	China, Peoples Rep. of	C, D, G, H
		Colombia	A, B
Bahamas	A, B	Congo, Dem. Rep of *(form. Zaire)*	E
Bahrain	G	Congo, Peoples Rep. of	C, E
Bangladesh	A, C, D	Costa Rica	A, B
Barbados	A, B, F, H	Cyprus	G
Belarus	C	Czech Republic	E
Belgium	A, C, E		
Belize	A, B, H	Denmark	C, K
Benin	D	Djibouti, Rep. of	C, E
Bermuda	A, B	Dominican Republic	A
Bolivia	A, C		
Botswana	C, D, H	Ecuador	A, B, C, D
Brazil	A, B, C	Egypt	C
Brunei	G	El Salvador	A, B, C, D, E, F, G, I, J, L
Bulgaria	F	England	A, C, H
Burkina Faso	B, E	Equatorial Guinea	C, E
Burma	C, D, F	Eritrea	C
Burundi	C, E	Ethiopia	C
		Fiji	I
Cameroon	C, E	Finland	C, F
Canada	B	France	E

Country	Plug Type	Country	Plug Type
Gabon	D, E	Lesotho	D
Gambia, The	G	Liberia	A, B
Germany, Fed. Rep. of	F	Luxembourg	F
Ghana	D, G		
Gibraltar	C, G	Macedonia	C, F
Greece	C, F	Madagascar	C, D, E, J, K
Greenland	C, K	Malawi	G
Grenada	G	Malaysia	G
Guatemala	A, B, G, H, I	Mali, Rep. of	C, E
Guinea	C, F, K	Malta	G
Guinea-Bissau	C	Mauritania	C
Guyana	A, H	Mauritius	G
		Mexico	A, B
Haiti	A, B, H	Monaco	C, D, E, F
Honduras	A	Morocco	C, E
Hong Kong	H	Mozambique	C, D, F
Hungary	C, F		
		Namibia	C
Iceland	B	Nepal	C, D
India	C, D, G	Netherlands	F
Indonesia	C, E, F	New Zealand	H
Ireland	G	Nicaragua	A
Israel	C, H	Niger	A, C, E
Italy	L	Nigeria	C, D, H
Ivory Coast	C, E	Northern Ireland	A, C, H
		Norway	C, F
Jamaica	A, B, C, D		
Japan	A, B, I	Oman	H
Jordan	C, F, G, L		
		Pakistan	B, C, D
Kazakstan	C, G, H	Palau	A, B
Kenya	G	Panama	A, B, I
Korea	C	Paraguay	C
Kuwait	C, G	Peru	A, C
		Philippines	A, B, C
Laos	A, B, C, E, F	Poland	C, E
Lebanon	A, B, C, D, G	Portugal	C, F

Country	Plug Type	Country	Plug Type
Qatar	D, G	Tahiti	A
		Taiwan	A, B
Romania	C, F	Tanzania	D, G
Russia	C	Thailand	A, B, C, D, E, G, J, K
Rwanda	C, J	Togo	C
		Trinidad and Tobago	A, B
Saudi Arabia	A, B, G	Tunisia	C, E
Scotland	A, C, H	Turkey	C, F
Senegal	C, D, E, K	Turkmenistan	B, F
Serbia-Montenegro	F		
Seychelles	D,	Uganda	G
Sierra Leone	D, G	Ukraine	C
Singapore	B, H	United Arab Emirates	C, D, G
Slovak Republic	E	Uruguay	C, F, I, L
Somalia	C	Uzbekistan	C, I
South Africa	D		
Spain	C, F	Venezuela	A, B, H
Sri Lanka	D		
Sudan	C, D	Wales	A, C, H
Suriname	C, F	Western Samoa	H
Swaziland	D		
Sweden	C, F	Yemen, Rep. of	A, D, G
Switzerland	C, E, J		
Syria	C	Zambia	C, D, G
		Zimbabwe	D, G
Tajikistan	C, I		

Plugs in Commercial Use

Type
A
Flat blade
attachment plug

Type
B
Flat blades with
round grounding pin

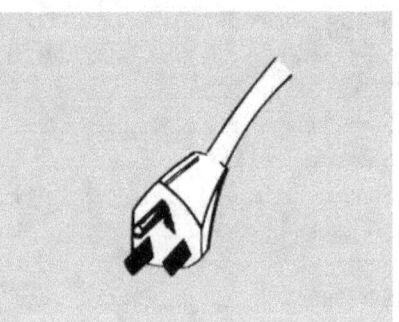

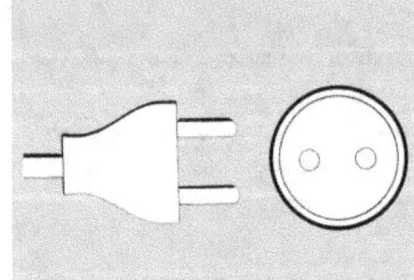

Type
C
Round pin
attachment plug

Type
D
Round pins
with ground

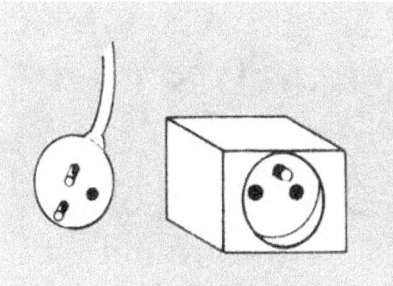

Type
E

Round pin plug and receptacle with male grounding pin

Type
F

"Schuko" plug and receptacle with side grounding contacts

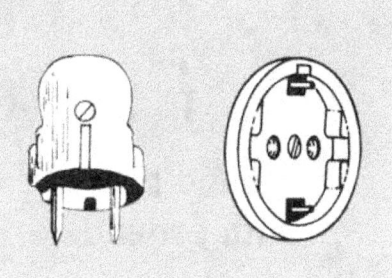

Type
G

Rectangular blade plug

Type
H

Oblique flat blades with ground

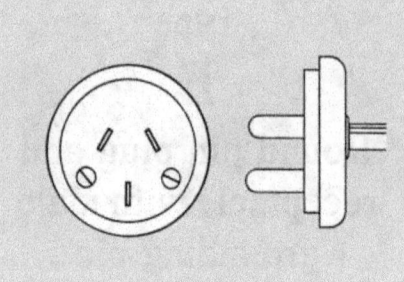

Type
I
Oblique flat blade
with ground

Type
J
Round pins
with ground

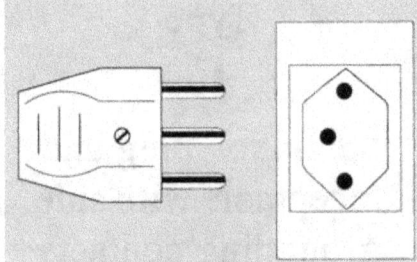

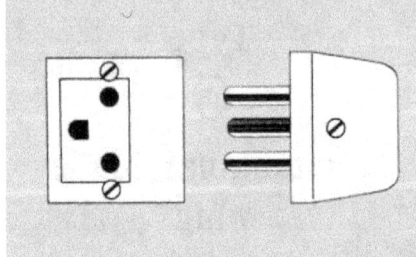

Type
K
Round pins
with ground

Type
L
Round pins
with ground

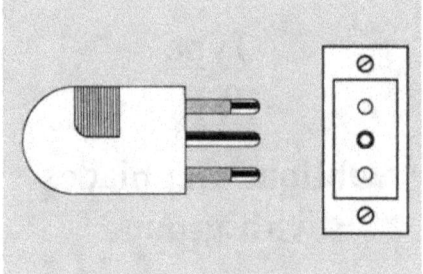

Electric Current Characteristics Abroad

Country or city	Type and frequency of current	Number of phases	Nominal voltage	Number of wires	Frequency stable enough for electric clocks?
Afghanistan	a.c. 50	1, 3	220/380	2, 4	yes
Albania	a.c. 50	1, 3	220/380	2, 4	no
Algeria	a.c. 50	1, 3	127/220 220/380	2, 4	yes
Angola[1, 2, 3]	a.c. 50	1, 3	220/380	2, 4	yes
Argentina	a.c. 50 d.c.	1, 3	220/380 220/440	2, 4 2, 3	yes yes
Australia[1, 2]	a.c. 50	1, 3	240/415	2, 3, 4	yes
Austria[1, 2]	a.c. 50	1, 3	220/380	3, 5	yes
Bahamas	a.c. 60	1, 3	120/240 120/208	2, 3, 4	yes
Bahrain[1, 2]	a.c. 50 d.c. 60	1, 3 1	230/400 110/115	2, 3, 4 3	yes yes
Bangladesh[1, 2, 3]	a.c. 50	1, 3	220/440	3, 4	no
Barbados[1, 2]	a.c. 50	1, 3	115/230 115/200	2, 3, 4	yes

1. The neutral wire of the secondary distribution system is grounded.
2. A grounding co0nductor is required in the electrical cord attached to appliances.
3. Voltage tolerance is plus or minus 4 to 9%.

Country or city	Type and frequency of current	Number of phases	Nominal voltage	Number of wires	Frequency stable enough for electric clocks?
Belarus	a.c. 50	1, 3	220/380	2, 4	yes
Belgium[1, 2]	a.c. 50	1, 3	220/400	2, 3, 4	yes
Belize[1, 3]	a.c. 60	1, 3	110/220 220/440	2, 3, 4	yes
Benin	a.c. 50	1, 3	220/380	2, 4	yes
Bermuda[1, 2, 3]	a.c. 60	1, 3	120/240 120/208	2, 3, 4	yes
Bolivia	a.c. 50	1, 3	110/220	2, 4	yes
Botswana	a.c. 50	1, 3	231/400	2, 4	yes
Brazil[1]					
Alagoinhas	a.c. 60	1, 3	127/220	2, 3, 4	yes
Americana	a.c. 60	1, 3	127/220	2, 3, 4	yes
Anapolis	a.c. 60	1, 3	220/380	2, 3, 4	yes
Aracaju	a.c. 60	1, 3	127/220	2, 3, 4	yes
Aracatuba	a.c. 60	1, 3	127/220	2, 3, 4	yes
Araraquara	a.c. 60	1, 3	127/220	2, 3, 4	yes
Bage	a.c. 60	1, 3	220/380	2, 3, 4	yes
Baixo Guandu	a.c. 60	1, 3	127/220	2, 3, 4	yes
Barbacena	a.c. 60	1, 3	127/220	2, 3, 4	yes
Barra Mansa	a.c. 60	1, 3	127/220	2, 3, 4	yes
Barretos	a.c. 60	1, 3	127/220	2, 3, 4	yes
Bauru	a.c. 60	1, 3	127/220	2, 3, 4	yes
Belem	a.c. 60	1, 3	127/220	2, 3, 4	yes
Belo Horizonte	a.c. 60	1, 3	127/220	2, 3, 4	yes
Blumenau	a.c. 60	1, 3	220/380	2, 3	yes
Boa Vista (Rio Branco)	a.c. 60	1, 3	127/220	2, 3, 4	yes
Botucatu	a.c. 60	1, 3	127/220	2, 3, 4	yes
Braganca	a.c. 60	1, 3	127/220	2, 3, 4	yes

1. The neutral wire of the secondary distribution system is grounded.
2. A grounding conductor is required in the electrical cord attached to appliances.
3. Voltage tolerance is plus or minus 4 to 9%.

Country or city	Type and frequency of current	Number of phases	Nominal voltage	Number of wires	Frequency stable enough for electric clocks?
Brazil,[1] *continued*					
Brazilia, D.F.	a.c. 60	1, 3	220/380	2, 3, 4	yes
Cachoeira	a.c. 60	1, 3	127/220	2, 3, 4	yes
Cachoeira do Itapemirim	a.c. 60	1, 3	127/220	2, 3, 4	yes
Campinas	a.c. 60	1, 3	127/220	2, 3, 4	yes
Campos	a.c. 60	1, 3	127/220	2, 3, 4	no
Caruaru	a.c. 60	1, 3	220/380	2, 3, 4	yes
Caxias do Sul	a.c. 60	1, 3	220/380	2, 3, 4	yes
Cel Fabriciano	a.c. 60	1, 3	110/220	2, 3	yes
Cidade Industrial (Betim)	a.c. 60	1, 3	127/220	2, 3, 4	yes
Colatina	a.c. 60	1, 3	127/220	2, 3, 4	yes
Corumba	a.c. 60	1, 3	127/220	2, 3, 4	yes
Curitiba	a.c. 60	1, 3	127/220	2, 3, 4	yes
Feira de Santana	a.c. 60	1, 3	127/220	2, 3, 4	yes
Florianopolis	a.c. 60	1, 3	220/380	2, 3, 4	yes
Fortaleza	a.c. 60	1, 3	220/380	2, 3, 4	yes
Franca (Sao Paulo)	a.c. 60	1, 3	127/220	2, 3, 4	yes
Goiania	a.c. 60	1, 3	220/380	2, 3, 4	yes
Goias	a.c. 60	1, 3	220/380	2, 3, 4	yes
Governador Valadares	a.c. 60	1, 3	127/220	2, 3, 4	yes
Ilheus	a.c. 60	1, 3	127/220	2, 3, 4	yes
Itabuana	a.c. 60	1, 3	127/220	2, 3, 4	yes
Itajai	a.c. 60	1, 3	220/380	2, 3	yes
Jequie	a.c. 60	1, 3	220/380	2, 3, 4	yes
Joao Pessoa	a.c. 60	1, 3	220/380	2, 3, 4	yes
Joinville	a.c. 60	1, 3	220/380	2, 3, 4	yes
Juiz de Fora	a.c. 60	1, 3	120/240	2, 3, 4	yes
Jundiai	a.c. 60	1, 3	220	2, 3	yes
Livramento	a.c. 60	1, 3	220/380	2, 3, 4	yes
Londrina	a.c. 60	1, 3	127/220	2, 3, 4	yes
Macapa	a.c. 60	1, 3	127/220	2, 3	yes
Maceio (Alagoas)	a.c. 60	1, 3	220/380	2, 3, 4	yes
Manaus	a.c. 60	1, 3	110/220	2, 3	yes

1. The neutral wire of the secondary distribution system is grounded.

Country or city	Type and frequency of current	Number of phases	Nominal voltage	Number of wires	Frequency stable enough for electric clocks?
Brazil,[1] *continued*					
Marilia	a.c. 60	1, 3	127/220	2, 3, 4	yes
Mossoro	a.c. 60	1, 3	220/380	2, 3, 4	yes
Natal	a.c. 60	1, 3	220/380	2, 3, 4	yes
(Rio Grando Norte)					
Niteroi	a.c. 60	1, 3	127/220	2, 3, 4	yes
Nova Friburgo	a.c. 60	1, 3	220/380	2, 3	yes
Olinda	a.c. 60	1, 3	220/380	2, 3, 4	yes
Ouro Preto	a.c. 50	1, 3	127/220	2, 3, 4	yes
Paranagua	a.c. 60	1, 3	127/220	2, 3	yes
Parnaiba	a.c. 60	1, 3	220/380	2, 3	yes
Paulista	a.c. 60	1, 3	127/220	2, 3, 4	yes
Pelotas	a.c. 60	1, 3	220/380	2, 3, 4	yes
Petropolis	a.c. 60	1, 3	127/220	2, 3, 4	yes
Piracicaba	a.c. 60	1, 3	127/220	2, 3, 4	yes
Ponta Grossa	a.c. 60	1, 3	127/220	2, 3	yes
Porto Alegre	a.c. 60	1, 3	127/220	2, 3, 4	yes
Porto Velho	a.c. 60	1, 3	127/220	2, 3	yes
Recife	a.c. 60	1, 3	220/380	2, 3, 4	yes
Ribeirao Preto	a.c. 60	1, 3	127/220	2, 3, 4	yes
Rio Branco	a.c. 60	1, 3	127/220	2, 3, 4	yes
Rio de Janeiro	a.c. 60	1, 3	127/220	2, 3, 4	yes
Salvador	a.c. 60	1, 3	127/220	2, 3, 4	yes
Santo André	a.c. 60	1, 3	127/220 220/380	2, 3	yes
Santos	a.c. 60	1, 3	127/220	2, 3, 4	yes
Sao Bernardo do Campo	a.c. 60	1, 3	220/380	2, 3	yes
Sao Caetano do Sul	a.c. 60	1, 3	115/230	2, 3	yes
Sao Luis	a.c. 60	1, 3	110/220	2, 3	yes
Sao Paulo	a.c. 60	1, 3	115/230	2, 3	yes
Sorocaba	a.c. 60	1, 3	127/220	2, 3, 4	yes
Teresina	a.c. 60	1, 3	110/220	2, 3	yes
Uberaba	a.c. 60	1, 3	127/220	2, 3, 4	yes
Vitoria	a.c. 60	1, 3	127/220	2, 3, 4	yes
Volta Redonda	a.c. 60	1, 3	125/216	2, 3, 4	yes

1. The neutral wire of the secondary distribution system is grounded.

Country or city	Type and frequency of current	Number of phases	Nominal voltage	Number of wires	Frequency stable enough for electric clocks?
Brunei[1, 2]	a.c. 50	1, 3	240/415	2, 4	yes
Bulgaria	a.c. 50	1, 3	220/380	2, 4	no
Burkina Faso	a.c. 50	1, 3	220/380	2, 4	no
Burma[1, 2, 3]	a.c. 50	1, 3	230/400	2, 4	no
Burundi[3]	a.c. 50	1, 3	220/380	2, 4	no
Cambodia	a.c. 50	1, 3	220/380	2, 3, 4	no
Cameroon	a.c. 50	1, 3	220/380	2, 4	yes
Canada[1]	a.c. 60	1, 3	120/240	3, 4	yes
Cape Verde[2]	a.c. 50	1, 3	220/380	2, 3, 4	no
Cayman Islands[1, 3]	a.c. 60	1, 3	120/240	2, 3	yes
Central African Republic[2, 3]	a.c. 50	1, 3	220/380	2, 4	yes
Chad	a.c. 50	1, 3	220/380	2, 4	no
Chile	a.c. 50	1, 3	220/380	2, 3, 4	yes
China, Peoples Republic of	a.c. 50	1, 3	220/380	2, 3, 4	no
Colombia	a.c. 60	1, 3	110/220 150/260	2, 3, 4	yes
Congo, Democratic Republic of the [1, 2] (formerly Zaire)	a.c. 50	1, 3	220/380	2, 3, 4	yes

1. The neutral wire of the secondary distribution system is grounded.
2. A grounding conductor is required in the electrical cord attached to appliances.
3. Voltage tolerance is plus or minus 4 to 9%.

Country or city	Type and frequency of current	Number of phases	Nominal voltage	Number of wires	Frequency stable enough for electric clocks?
Congo, Republic of [1,2,3]					
	a.c. 50	1, 3	220/380	2, 4	no
Costa Rica	a.c. 60	1, 3	120/240	2, 3, 4	yes
Cyprus [1,2]	a.c. 50	1, 3	240/415	2, 4	yes
Czech Republic	a.c. 50	1, 3	220/380	2, 3, 4	yes
Denmark	a.c. 50	1, 3	220/380	2, 3, 4	yes
Djibouti, Republic of					
	a.c. 50	1, 3	220/380	2, 4	yes
Dominican Republic					
	a.c. 60	1, 3	110/220	2, 3	yes
Ecuador [1]	a.c. 60	1, 3	120/208 127/220	2, 3, 4	yes
Egypt	a.c. 50	1, 3	220/380	2, 3, 4	no
El Salvador [1]	a.c. 60	1, 3	115/230	2, 3	yes
England (See United Kingdom)					
Eritrea	a.c. 50	1, 3	220/380	2, 4	yes
Ethiopia	a.c. 50	1, 3	220/380	2, 4	yes
Fiji [3]	a.c. 50	1, 3	240/415	2, 3, 4	yes
Finland	a.c. 50	1, 3	230/400	2, 4, 5	yes
France	a.c. 50	1, 3	220/380	2, 4	yes

---

1. The neutral wire of the secondary distribution system is grounded.
2. A grounding conductor is required in the electrical cord attached to appliances.
3. Voltage tolerance is plus or minus 4 to 9%.

Country or city	Type and frequency of current	Number of phases	Nominal voltage	Number of wires	Frequency stable enough for electric clocks?
Gabon[1,2]	a.c. 50	1, 3	220/380	2, 4	yes
Gambia, The[1,2]	a.c. 50	1, 3	220/380	2, 4	no
Germany, Federal Republic of [1,2,3]					
	a.c. 50	1, 3	230/400	2, 4	yes
Ghana	a.c. 50	1, 3	240/415	2, 4	no
Gibraltar	a.c. 50	1, 3	240/415	2, 4	yes
Great Britain (See United Kingdom)					
Greece	a.c. 50	1, 3	220/380	2, 4	yes
Greenland	a.c. 50	1, 3	220/380	2, 3, 4	yes
Grenada[1,2,3]	a.c. 50	1, 3	230/400	2, 4	no
Guatemala	a.c. 60	1, 3	120/240	2, 3, 4	yes
Guinea	a.c 50	1, 3	220/380	2, 3, 4	no
Guinea-Bissau	a.c. 50	1, 3	220/380	2, 3, 4	no
Guyana[1,2]	a.c. 50	1, 3	110/220	2, 3, 4	yes
Haiti	a.c.60	1, 3	110/220	2, 3, 4	no
Honduras	a.c. 60	1, 3	110/220	2, 3	no
Hong Kong	a.c. 50	1, 3	202/415	3, 4	yes
Hungary[2,3]	a.c. 50	1, 3	220/380	2, 3, 4	yes
Iceland	a.c. 50	1, 3	220/380	2, 3, 4	yes

1. The neutral wire of the secondary distribution system is grounded.
2. A grounding conductor is required in the electrical cord attached to appliances.
3. Voltage tolerance is plus or minus 4 to 9%.

Country or city	Type and frequency of current	Number of phases	Nominal voltage	Number of wires	Frequency stable enough for electric clocks?
India[3]	a.c. 50	1, 3	230/400	2, 4	yes
Indonesia[1]					
Bandjarmasin	a.c. 50	1, 3	127/220	2, 4	no
Bandung	a.c. 50	1, 3	220/380	2, 4	yes
Bogor	a.c. 50	1, 3	220/380	2, 4	yes
Cilacap	a.c. 50	1, 3	220/380	2, 4	yes
Cirebon	a.c. 50	1, 3	220/380	2, 4	yes
Jakarta	a.c. 50	1, 3	220/380	2, 4	yes
Malang	a.c. 50	1, 3	220/380	2, 4	yes
Medan	a.c. 50	1, 3	127/220	2, 4	yes
Padang	a.c. 50	1, 3	127/220	2, 4	no
Palembang	a.c. 50	1, 3	127/220	2, 4	yes
Semarang	a.c. 50	1, 3	220/380	2, 4	yes
Sukabumi	a.c. 50	1, 3	220/380	2, 4	yes
Surabaya	a.c. 50	1, 3	220/380	2, 4	yes
Surakarta	a.c. 50	1, 3	220/380	2, 4	yes
Ujungpandang	a.c. 50	1, 3	127/220	2, 4	no
Yogyakarta	a.c. 50	1, 3	220/380	2, 4	yes
Ireland[1,2,3]	a.c. 50	1, 3	220/380	2, 4	yes
Israel[1,2,4]	a.c. 50	1, 3	220/380	2, 4	yes
Italy[1,2,4]					
Ancona	a.c. 50	1, 3	127/220 220/380	2, 4	yes
Bari	a.c. 50	1, 3	220/380	2, 4	yes
Bologna	a.c. 50	1, 3	127/220 220/380	2, 4	yes
Brindisi	a.c. 50	1, 3	220/380	2, 4	yes
Cagliari	a.c. 50	1, 3	220/380	2, 4	yes
Catania	a.c. 50	1, 3	220/380	2, 4	yes
Como	a.c. 50	1, 3	127/220 220/380	2, 4	yes

1. The neutral wire of the secondary distribution system is grounded.
2. A grounding conductor is required in the electrical cord attached to appliances.
3. Voltage tolerance is plus or minus 4 to 9%.
4. Voltage tolerance is plus or minus 10%.

Country or city	Type and frequency of current	Number of phases	Nominal voltage	Number of wires	Frequency stable enough for electric clocks?
Italy,[1,2,4] *continued*					
Cremona	a.c. 50	1, 3	127/220 220/380	2, 4	yes
Florence	a.c. 50	1, 3	220/380	2, 4	yes
Genoa	a.c. 50	1, 3	127/220 220/380	2, 4	yes
La Spezia	a.c. 50	1, 3	220/380	2, 4	yes
Latina	a.c. 50	1, 3	127/220 220/380	2, 4	yes
Leghorn	a.c. 50	1, 3	220/380	2, 4	yes
Milan	a.c. 50	1, 3	127/220 220/380	2, 4	yes
Naples	a.c. 50	1, 3	220/380	2, 4	yes
Palermo	a.c. 50	1, 3	220/380	2, 4	yes
Perugia	a.c. 50	1, 3	127/220 220/380	2, 4	yes
Pescara and Chieti	a.c 50	1, 3	127/220 220/380	2, 4	yes
Pisa	a.c. 50	1, 3	127/220 220/380	2, 4	yes
Ragusa	a.c. 50	1, 3	220/380	2, 4	yes
Rome	a.c. 50	1, 3	127/220 220/380	2, 4	yes
Sassari	a.c. 50	1, 3	220/380	2, 4	yes
Siena	a.c. 50	1, 3	220/380	2, 4	yes
Siracusa	a.c. 50	1, 3	220/380	2, 4	yes
Taranto	a.c. 50	1, 3	220/380	2, 4	yes
Trieste	a.c. 50	1, 3	127/220 220/380	2, 4	yes
Turin	a.c. 50	1, 3	220/380	2, 4	yes
Udine	a.c. 50	1, 3	127/220 220/380	2, 4	yes
Venice	a.c. 50	1, 3	127/220 220/380	2, 4	yes
Verona	a.c. 50	1, 3	127/220 220/380	2, 4	yes

1. The neutral wire of the secondary distribution system is grounded.
2. A grounding conductor is required in the electrical cord attached to appliances.
4. Voltage tolerance is plus or minus 10%.

Country or city	Type and frequency of current	Number of phases	Nominal voltage	Number of wires	Frequency stable enough for electric clocks?
Ivory Coast	a.c. 50	1, 3	220/380	3, 4	yes
Jamaica[1,3]	a.c. 50	1, 3	110/220	2, 3, 4	yes
Japan[1]	a.c. 50	1, 3	100/200	2, 3	yes
Jordan[1,2]	a.c. 50	1, 3	220/380	2, 3, 4	yes
Kazakstan	a.c. 50	1, 3	220/380	2, 3, 4	yes
Kenya[2,3]	a.c. 50	1, 3	240/415	2, 4	no
Korea, Republic of [1,2,3]	a.c. 60	1, 3	220/380	2, 4	yes
Kuwait[3]	a.c. 50	1, 3	240/415	2, 4	yes
Laos	a.c. 50	1, 3	220/380	2, 4	yes
Lebanon[1]					
Aley	a.c. 50	1, 3	110/190 220/380	2, 4	no
Beirut	a.c. 50	1, 3	110/190 220/380	2, 4	no
Bhamdoun	a.c. 50	1, 3	110/190 220/380	2, 4	no
Brummana	a.c. 50	1, 3	110/190 220/380	2, 4	no
Chtaure	a.c. 50	1, 3	220/380	2, 4	no
Dhour el Choueir	a.c. 50	1, 3	220/380	2, 4	no
Sidon	a.c. 50	1, 3	220/380	2, 4	no
Sofar	a.c. 50	1, 3	220/380	2, 4	no
Tripoli	a.c. 50	1, 3	110/190 220/380	2, 4	no
Tyre	a.c. 50	1, 3	110/190 220/380	2, 4	no
Zahleh	a.c. 50	1, 3	220/380	2, 4	no

1. The neutral wire of the secondary distribution system is grounded.
2. A grounding conductor is required in the electrical cord attached to appliances.
3. Voltage tolerance is plus or minus 4 to 9%.

Country or city	Type and frequency of current	Number of phases	Nominal voltage	Number of wires	Frequency stable enough for electric clocks?
Lesotho[1, 2]	a.c. 50	1, 3	220/380	2, 4	yes
Liberia[1]	a.c. 60	1, 3	120/240	2, 3, 4	no
Luxembourg[1, 2]	a.c. 50	1, 3	230/400	2, 4,	yes
Macedonia	a.c. 50	1, 3	220/380	2, 4	yes
Madagascar[1, 2]	a.c. 50	1, 3	127/220 220/380	2, 3, 4	yes
Malawi[3]	a.c. 50	1, 3	230/400	3, 4	no
Malaysia[1, 2]	a.c. 50	1, 3	240/415	2, 3	yes
Mali, Republic of [1, 2]	a.c. 50	1, 3	220/380	3, 4	no
Malta[1, 2]	a.c. 50	1, 3	240/415	2, 4	yes
Mauritania[1, 2, 5]	a.c. 50	1, 3	220/380	2, 3	no
Mauritius[1, 2]	a.c. 50	1, 3	230/400	2, 4	yes
Mexico[1]	a.c. 60	1, 3	127/220	2, 3, 4	yes
Monaco	a.c. 50	1, 3	127/220 220/380	2, 4	yes
Morocco[1, 2]					
Agadir	a.c. 50	1, 3	127/220 220/380	2, 4	yes
Asilah	a.c. 50	1, 3	127/220	2, 4	yes
Beni-Mellal	a.c. 50	1, 3	127/220 220/380	2, 4	yes
Berrechid	a.c. 50	1, 3	127/220	2, 4	yes

1. The neutral wire of the secondary distribution system is grounded.
2. A grounding conductor is required in the electrical cord attached to appliances.
3. Voltage tolerance is plus or minus 4 to 9%.
5. Voltage tolerance is plus or minus 20 to 30%.

Country or city	Type and frequency of current	Number of phases	Nominal voltage	Number of wires	Frequency stable enough for electric clocks?
Morocco,[1,2] *continued*					
Chaouen	a.c. 50	1, 3	127/220	2, 4	yes
Casablanca	a.c. 50	1, 3	127/220	2, 4	yes
El-Jadida	a.c. 50	1, 3	127/220	2, 4	yes
El-Hoceima	a.c. 50	1, 3	220/380	2, 4	yes
Fes	a.c. 50	1, 3	127/220	2, 4	yes
Ksar-Es-Souk	a.c. 50	1, 3	127/220	2, 4	yes
Kasba-Tadla	a.c. 50	1, 3	127/220	2, 4	yes
Khemisset	a.c. 50	1, 3	220/380	2, 4	yes
Khenifra	a.c. 50	1, 3	220/380	2, 4	yes
Khouribga	a.c. 50	1, 3	127/220	2, 4	yes
Ksar-El-Kebir	a.c. 50	1, 3	127/220	2, 4	yes
Larache	a.c. 50	1, 3	127/220	2, 4	yes
Mohammedia	a.c. 50	1, 3	127/220	2, 4	yes
Marrakech	a.c. 50	1, 3	127/220	2, 4	yes
Meknes	a.c. 50	1, 3	127/220	2, 4	yes
Midelt	a.c. 50	1, 3	127/220	2, 4	yes
Nador	a.c. 50	1, 3	127/220	2, 4	yes
Ouarzazete	a.c. 50	1, 3	127/220	2, 4	yes
Oued-Zem	a.c. 50	1, 3	127/220 220/380	2, 4	yes
Ouezzane	a.c. 50	1, 3	127/220	2, 4	yes
Rabat	a.c. 50	1, 3	127/220	2, 4	yes
Safi	a.c. 50	1, 3	127/220	2, 4	yes
Sefrou	a.c. 50	1, 3	127/220	2, 4	yes
Sidi Kacem	a.c. 50	1, 3	127/220 220/380	2, 4	yes
Sidi Slimane	a.c. 50	1, 3	127/220 220/380	2, 4	yes
Souk-El-Arba Gharb	a.c. 50	1, 3	127/220 220/380	2, 4	yes
Settat	a.c. 50	1, 3	127/220	2, 4	yes
Taza	a.c. 50	1, 3	127/220	2, 4	yes
Taroudant	a.c. 50	1, 3	127/220	2, 4	yes
Tiznit	a.c. 50	1, 3	127/220	2, 4	yes
Tangier	a.c. 50	1, 3	127/220	2, 4	yes
Tetouan	a.c. 50	1, 3	127/220	2, 4	yes
Youssoufia	a.c. 50	1, 3	127/220	2, 4	yes

1. The neutral wire of the secondary distribution system is grounded.
2. A grounding conductor is required in the electrical cord attached to appliances.

Country or city	Type and frequency of current	Number of phases	Nominal voltage	Number of wires	Frequency stable enough for electric clocks?
Mozambique[2]	a.c. 50	1, 3	220/380	2, 4	yes
Namibia[1,2]	a.c. 50	1, 3	220/380	2, 4	yes
Nepal[1]	a.c. 50	1, 3	220/380	2, 4	no
Netherlands[1]	a.c. 50	1,3	220/380	2, 4	yes
New Zealand[1,2]	a.c. 50	1, 3	230/400	2, 3, 4	yes
Nicaragua	a.c. 60	1, 3	120/240	2, 3, 4	yes
Niger	a.c. 50	1, 3	220/380	2, 4	no
Nigeria[1]	a.c. 50	1, 3	220/380	2, 4	yes
Northern Ireland (See United Kingdom)					
Norway	a.c. 50	1, 3	220/380	2, 4	yes
Oman[2]	a.c. 50	1, 3	240/415	2, 4	yes
Pakistan[1]	a.c. 50	1, 3	230/400	3	no
Palau	a.c. 60	1, 3	120/240	4	no
Panama	a.c. 60	1, 3	120/240	2, 4	yes
Paraguay	a.c. 50	1, 3	220/380	2, 4	yes
Peru	a.c. 60	1, 3	220/380	2, 4	yes
Philippines[1,2]	a.c. 60	1, 3	125/216	2, 4	yes
Poland	a.c. 50	1, 3	220/380	3, 4	yes
Portugal[1]	a.c. 50	1, 3	220/380	2, 3, 4	yes

1. The neutral wire of the secondary distribution system is grounded.
2. A grounding conductor is required in the electrical cord attached to appliances.

Country or city	Type and frequency of current	Number of phases	Nominal voltage	Number of wires	Frequency stable enough for electric clocks?
Qatar	a.c. 50	1, 3	240/415	2, 3, 4	yes
Romania[3]	a.c. 50	1, 3	220/380	2, 4	no
Russia	a.c. 50	1, 3	220/380	2, 4	yes
Rwanda	a.c. 50	1, 3	220/380	2, 4	yes
Saudi Arabia[3]	a.c. 60	1, 3	127/220	2, 4	yes
Scotland *(See United Kingdom)*					
Senegal[1,3]	a.c. 50	1, 3	127/220	2, 3, 4	no
Serbia-Montenegro					
	a.c. 50	1, 3	220/380	3, 4, 5	yes
Seychelles	a.c. 50	1, 3	240/450	2, 4	yes
Sierra Leone	a.c. 50	1, 3	230/400	2, 4	no
Singapore[1]	a.c. 50	1, 3	230/400	2, 3	yes
Slovak Republic[1]					
	a.c 50	1, 3	220/380	2, 4	yes
Somalia					
Berbera	a.c. 50	1, 3	230	2, 3	yes
Brava	a.c. 50	1, 3	220/440	2, 4	yes
Chisimaio	a.c. 50	1, 3	220	2, 3	no
Hargeisa	a.c. 50	1, 3	220	2, 3	yes
Merca	a.c. 50	1, 3	110/220	2, 4	no
Mogadishu	a.c. 50	1, 3	220/380	2, 4	no
South Africa[1,2,3]					
Alberton	a.c. 50	1, 3	220/380	2, 3, 4	yes

---

1. The neutral wire of the secondary distribution system is grounded.
2. A grounding conductor is required in the electrical cord attached to appliances.
3. Voltage tolerance is plus or minus 4 to 9%.

Country or city	Type and frequency of current	Number of phases	Nominal voltage	Number of wires	Frequency stable enough for electric clocks?
South Africa,[1,2,3] *continued*					
Beaufort West	a.c. 50	1, 3	230/400	2, 4	yes
Benoni	a.c. 50	1, 3	230/400	2, 3, 4	yes
Bethlehem	a.c. 50	1, 3	220/380	2, 4	yes
Bloemfontein	a.c. 50	1, 3	220/380	2, 4	yes
Boksburg	a.c. 50	1, 3	230/400	2, 4	yes
Brakpan	a.c. 50	1, 3	220/380	2, 3, 4	yes
Caledon	a.c. 50	1, 3	220/380	2, 4	yes
Cape Town	a.c. 50	1, 3	220/380	2, 4	yes
Carltonville	a.c. 50	1, 3	220/380	2, 4	yes
Cradock	a.c. 50	1, 3	230/400	2, 4	n.a.
De Aar	a.c. 50	1, 3	220/380	2, 4	yes
Durban	a.c. 50	1, 3	220/380	2, 4	yes
East London	a.c. 50	1, 3	220/380	2, 4	yes
Germiston	a.c. 50	1, 3	230/400	2, 3, 4	yes
Grahamstad	a.c. 50	1, 3	250/430	2, 4	yes
Johannesburg	a.c. 50	1, 3	220/380	2, 3, 4	yes
	d.c.		230/460	2, 3	
Kimberley	a.c. 50	1, 3	220/380	2, 3, 4	yes
King Williams	a.c. 50	1, 3	220/380	2, 3, 4	yes
			250/433		
Klerksdorp	a.c. 50	1, 3	230/400	2, 3, 4	yes
Kroonstad	a.c. 50	1, 3	230/400	2, 3, 4	yes
Krugersdorp	a.c. 50	1, 3	220/380	2, 4	yes
Malmesbury	a.c. 50	1, 3	220/380	2, 4	yes
Ladysmith, N.	a.c. 50	1, 3	220/380	2, 4	yes
Oudtshoorn	a.c. 50	1, 3	220/380	2, 4	yes
Paarl	a.c. 50	1, 3	230/400	2, 4	yes
Parys	a.c. 50	1, 3	220/380	2, 3, 4	yes
Pietermaritzburg	a.c. 50	1, 3	220/380	2, 4	yes
Port Elizabeth	a.c. 50	1, 3	250/433	2, 4	yes
Pretoria	a.c. 50	1, 3	240/415	2, 3, 4	yes
Queenstown	a.c. 50	1, 3	220/380	2, 4	yes
Robertson	a.c. 50	1, 3	220/380	2, 4	yes
Roodepoort	a.c. 50	1, 3	230/400	2, 4	yes

1. The neutral wire of the secondary distribution system is grounded.
2. A grounding conductor is required in the electrical cord attached to appliances.
3. Voltage tolerance is plus or minus 4 to 9%.

Country or city	Type and frequency of current	Number of phases	Nominal voltage	Number of wires	Frequency stable enough for electric clocks?
South Africa,[1,2,3] *continued*					
Rustenburg	a.c. 50	1, 3	220/380	2, 4	yes
Senekal	a.c. 50	1, 3	220/380	2, 3, 4	yes
Somerset West	a.c. 50	1, 3	230/400	2, 4	yes
Springs	a.c. 50	1, 3	220/380 230/400	2, 3, 4	yes
Stellenbosch	a.c. 50	3	220/380	4	yes
Tulbagh	a.c. 50	1, 3	220/380	2, 4	yes
Uitenhage	a.c. 50	1, 3	220/380	2, 4	yes
Umtata	a.c. 50	1, 3	230/400	2, 3, 4	yes
Umkomaas	a.c. 50	1, 3	220/380	2, 4	yes
Upington	a.c. 50	1, 3	230/400	2, 4	yes
Vereeniging	a.c. 50	1, 3	220/380	2, 4	yes
Virginia	a.c. 50	1, 3	230/400	2, 4	yes
Vryheid	a.c. 50	1, 3	230/400	2, 3, 4	yes
Walvis Bay	a.c. 50	1, 3	230/400	2, 3, 4	yes
Welkom	a.c. 50	1, 3	220/380	2, 4	yes
Wellington	a.c. 50	1, 3	230/400	2, 4	yes
Worcester	a.c. 50	1, 3	230/400	2, 4	yes
Spain[1]	a.c. 50	1, 3	220/380	2, 3, 4	yes
Sri Lanka[1,3]	a.c. 50	1, 3	230/400	2, 4	yes
Sudan[1]	a.c. 50	1, 3	240/415	2, 4	yes
Suriname	a.c. 60	1, 3	127/220	2, 3, 4	yes
Swaziland	a.c. 50	1, 3	230/400	2, 4	yes
Sweden[1,2]	a.c. 50	1, 3	230/400	2, 3, 4, 5	yes
Switzerland[1,2]	a.c. 50	1, 3	220/380	2, 3, 4	yes
Syria	a.c. 50	1, 3	220/380	2, 3	no

1. The neutral wire of the secondary distribution system is grounded.
2. A grounding conductor is required in the electrical cord attached to appliances.
3. Voltage tolerance is plus or minus 4 to 9%.

Country or city	Type and frequency of current	Number of phases	Nominal voltage	Number of wires	Frequency stable enough for electric clocks?
Tahiti	a.c. 60	1, 3	127/220	2, 3, 4	no
Taiwan[1]	a.c. 60	1, 3	110/220	2, 3, 4	yes
Tajikistan	a.c. 50	1, 3	220/380	2, 3	no
Tanzania[1, 2, 3]	a.c. 50	1, 3	220/380	2, 4	yes
Thailand[3]	a.c. 50	1, 3	220/380	2, 4	yes
Togo	a.c. 50	1, 3	127/220 220/380	2, 4	yes
Trinidad and Tobago[3]	a.c. 60	1, 3	115/230 230/400	2, 3, 4	yes
Tunisia[1, 2, 3]	a.c. 50	1, 3	220/380 220/380	2, 4	yes
Turkey[1]	a.c. 50	1, 3	220/380	2, 3, 4	yes
Turkmenistan	a.c. 50	1, 3	220/380	2, 3	yes
Uganda[1, 2]	a.c. 50	1, 3	240/415	2, 4	no
Ukraine[1]	a.c. 50	1, 3	220/380	2, 4	yes
United Arab Emirates	a.c. 50	1, 3	220/380	2, 4	yes
United Kingdom[1, 2, 3]					
England	a.c. 50	1	230/415	2, 4	yes
Northern Ireland	a.c. 50	1, 3	230/415	2, 4	yes
Scotland	a.c. 50	1, 3	230/415	2, 4	yes
Wales	a.c. 50	1, 3	230/415	2, 4	yes

1. The neutral wire of the secondary distribution system is grounded.
2. A grounding conductor is required in the electrical cord attached to appliances.
3. Voltage tolerance is plus or minus 4 to 9%.

Country or city	Type and frequency of current	Number of phases	Nominal voltage	Number of wires	Frequency stable enough for electric clocks?
Uruguay[2, 3, 6]	a.c. 50	1, 3	220/380	2, 4	yes
Uzbekistan	a.c. 50	1, 3	220/380	2, 4	yes
Venezuela	a.c. 60	1, 3	120/240	2, 3, 4	yes
Vietnam					
Ban Me Thuot (Sic)	a.c. 50	1, 3	220/380	2, 4	no
Can Tho	a.c. 50	1, 3	127/220 220/380	2, 4	no
Dalat	a.c. 50	1, 3	120/208 220/380	2, 4	no
Da Nang	a.c. 50	1, 3	127/220	2, 4	no
Hanoi	a.c. 50	1, 3	127/220 220/380	2, 4	no
Hue	a.c. 50	1, 3	127/220	2, 4	no
Khanh Hung (Soc Trang)	a.c. 50	1, 3	220/380	2, 4	no
Nha Trang	a.c. 50	1, 3	127/220	2, 4	no
Saigon	a.c. 50	1, 3	120/208 220/380	2, 4	no
Western Samoa	a.c. 50	1, 3	230/400	2, 3, 4	yes
Wales (See United Kingdom)					
Yemen, Republic of	a.c. 50	1, 3	220/380	2, 4	no
Zambia[1, 2, 4]	a.c. 50	1, 3	220/380	2, 4	yes
Zimbabwe[4]	a.c. 50	1, 3	220/380	2, 3, 4	yes

---

1. The neutral wire of the secondary distribution system is grounded.
2. A grounding conductor is required in the electrical cord attached to appliances.
3. Voltage tolerance is plus or minus 4 to 9%.
4. Voltage tolerance is plus or minus 10%.
6. Voltage tolerance is plus or minus 4.5 to 20.5%.